The Best Bits of Physics

by Alasdair C Shaw

First published March 2013
ISBN: 1482797259
ISBN-13: 978-1482797251

Foreword

Physics is a hard subject. Everyone knows that. Tell someone that you are studying physics and they will be impressed. This will probably be followed by a comment along the lines of "I could never do that at school". It is still one of the most demanding GCSEs and A-Levels.

This book is an attempt to serve up some of the best bits of physics. It should give you an understanding of the key concepts in modern physics. Along the way I hope you will be convinced that the heart of physics isn't all that hard after all. Perhaps you'll even be able to explain it to your friends and family...

Contents

Key Theories

Some of the most important bits of physics.

Conservation of Energy

The Principle of Conservation of Energy is one of the most fundamental of all physical laws. It underpins the vast majority of science and earns itself the title of being a 'First Principle'.

Simply put energy is always conserved. That means the total amount of energy in the universe is always the same. Another way of putting it is that energy is neither created nor destroyed.

That doesn't mean that the energy of a particular thing stays the same. If that were true you could run all day and never need to eat! Energy can be transferred from one thing to another. For instance the energy in my chocolate bar is about to be transferred into me. Yummy.

Energy can also be transformed from one type (called a form) to another. But what are these forms? You probably weren't asking, but I'll tell you anyway. They split into two broad groups – 'doing energy' and 'potential to do something energy'. OK, not their real names but close enough. Within these there are a whole host of energy forms – heat, light, sound, movement (bonus marks for calling it kinetic), gravitational potential, elastic potential, nuclear potential and magnetic potential to name but a few. No prizes for guessing which fall into which group.

Right, example time. A careless worker drops a hammer from the roof of a building and it falls straight to the floor. Just before he let it go it had 120J of gravitational potential energy (GPE for short). As it falls it loses GPE – it is losing height. At the same time it gains kinetic energy (KE) – it speeds up.

A fraction of a second before it hits the ground it has lost all its GPE – it has no height. It also has its maximum KE – it is travelling its fastest. Now, we know it started with 120J of energy. The Principle of Conservation of Energy tells us that the total energy must be the same. So it has lost 120J of GPE and gained 120J of KE.

Hang on, isn't that too simple? Well, to be honest it is a bit of a simplification. I have assumed that there is no air resistance. However for something as small and heavy as a hammer falling such a short distance it is a reasonably good assumption.

The changes of energy form are caused by forces. In the example above the main change (GPE to KE) is caused by the weight of the hammer. The small change I ignored (KE to heat and sound) is caused by air resistance.

In nearly all real situations there are resistive forces. The most common of these are friction and air resistance. These convert kinetic energy into heat and sound – try rubbing your hands together. This means that energy is always being 'lost' by moving objects. It hasn't been destroyed; it has just transferred into the surroundings. Over time this energy becomes so spread out that it is effectively useless.

Newton's Laws of Motion

These are some of the most basic statements about how things move. They replaced the teachings of Aristotle on the subject, which had held sway for thousands of years.

Newton's 1st Law of Motion states that "unless there is an unbalanced force acting upon it an object at rest will remain at rest and an object in motion will continue to move in the same direction at the same speed". OK, so he wrote it in Latin. Put simply it means that things keep doing what they were doing unless something pushes them. Well that's obvious isn't it? Perhaps now.

Newton's 2nd Law of Motion states that "the acceleration of an object is directly proportional to the force applied and inversely proportional to its mass". In other words the more you push it the quicker it speeds up and the more mass it has the slower it speeds up. This leads to one of the equations people seem to remember from physics - F=ma.

While F=ma works for most everyday situations it assumes that the mass of the object remains pretty much constant. Sometimes, however, physics is rocket science; for instance when launching a rocket. Due to the huge amount of fuel being used the rocket loses a lot of mass as it goes. This means we'd have to use the more general form of the law - force is rate of change of momentum. Momentum is simply mass times speed (velocity to be precise but don't worry about that now).

Newton's 3rd Law of Motion is one everyone has heard of, though perhaps not quite understood - "every action has an equal and opposite reaction". This means that if you push on something it pushes back on you with the same force.

Going back to our rocket we can now see how it works. The rocket applies a large force to push the burnt fuel out the back. According to Newton's 3rd Law the burnt fuel applies an equal sized force forwards on the rocket. This is why in rockets and spaceships the fuel is often called reaction mass.

Newton went on to formulate his Law of Universal Gravitation. The 'universal' bit refers to him realising that everything is attracted to everything else by gravity.

It states that "the force between two objects is in proportion to the product of their masses and inversely proportional to the square of the distance between them". Quite a mouthful. Basically the more mass each object has the stronger the force due to gravity. The further apart they are, however, the weaker the force due to gravity.

The gravitational forces on the two objects form a Third Law pair. My weight in the Earth's gravitational field is the same size as and in the opposite direction to the Earth's weight in my gravitational field.

Thermodynamics

Thermodynamics is the study of the movement of heat. When it started it was a very contentious area of science. Indeed some of its proponents committed suicide, partly blamed on the rest of the world not believing their theories.

First of all we need to make something clear. We don't always use scientific words correctly in normal English. Often this is because we don't fully understand them. What if I were to say something can be cold but have a lot of heat? Sounds wrong but it is perfectly possible!

Temperature is a measure of how hot or cold things are. Heat is the amount of internal energy they have. A lake at 5°C has more heat than a cup of coffee at 80°C because it is so much larger.

The first thing we looked at in this book was the Principle of Conservation of Energy. It sounds pretty simple, but it didn't quite work until the 19th century.

In the 1850s Lord Kelvin was able to prove that heat was a form of energy. He measured the temperature increase in water falling over a waterfall and related it to the loss in GPE of the water. This became the 1st Law of Thermodynamics - heat is a form of energy (or in the words of Flanders and Swan "heat is work and work is heat").

The 2nd Law of Thermodynamics followed. This stated that heat naturally flows from hot objects to cooler objects. To make it go the other way you have to do work. This is why a fridge needs to be plugged in - the electricity is required to drive the pump.

Next came the 0th Law of Thermodynamics. Yep, that's right. It was deemed to be so basic that it had to go before the 1st Law. It simply says that if A is at the same temperature as B and C is at the same temperature as B then C is at the same temperature as A. That sounds obvious, but it was a revelation.

One example of why it may not be immediately obvious is a bicycle left out in the cold. The rubber handle is at the same temperature as the air. The metal frame is at the same temperature as the air. If you put your hands on them the metal frame will feel a lot colder than the rubber handle. This is actually because the metal conducts the heat of your hand away faster.

Finally came the 3rd Law of Thermodynamics. This states that the amount of disorder in a system naturally increases. If you want to locally make something more ordered it is at the expense of extra disorder elsewhere. Think of a teenager's bedroom. Left to its own devices this will become progressively more messy. To tidy it up will take a lot of work.

Remember what happens in a system with resistive forces? The energy spreads out and becomes less useable. This is increasing disorder.

A great word to use is entropy. This is simply a measure of how disordered a system is. In other words entropy always tends to increase. This runs in the face of other physical laws. Apart from this all physical laws allow things to run equally well in reverse; it is the only thing that says a whole load of fragments of pottery can't be jiggled together by vibrations in the ground and then jump back onto the table as a whole coffee cup. The idea that entropy always increases leads to ideas about why we perceive time to flow in a particular direction.

Absolute Zero

What is the coldest possible temperature? Your first thought is probably 'there isn't one, you could just keep getting colder'. What if I were to tell you it was -273°C? Don't believe me?

All matter is made of particles. If you want to get really detailed it is made of atoms and molecules, but it is fine to just say particles.

In a solid the particles are fixed into a regular pattern and just vibrate. In liquids they move around slowly. In gases they move around rapidly. The key thing is that the particles are always moving.

Temperature is a measure of the kinetic energy of the particles in a material. As the particles slow down it gets colder. When they stop moving it has reached the coldest it can get, what we call absolute zero. It can't get any colder because the particles can't move slower than not moving.

This happens at -273°C. It doesn't matter what the material is or any other factor, it always happens at -273°C. This is so fundamental that a new temperature scale was invented by Lord Kelvin (yes, him again). Called the 'Kelvin Scale' or 'absolute temperature' it has absolute zero as 0K (that is zero kelvin) and the standard melting point of water at 273K. In other words all you need to do to get from °C to K is add 273.

Electromagnetism

I am sure you've heard of electricity and of magnetism. But what about electromagnetism? One thing that keeps happening in physics is that two apparently separate things turn out to be linked. Often we don't bother to make up a completely new word; instead we stick the two existing words together.

When a current runs through a wire a magnetic field is created. Coil the wire up and you add all the little magnetic fields into one big magnetic field. You have an electromagnet!

Indeed we now understand that all magnets are really electromagnets. A piece of iron is only magnetic if its domains line up. The domains are regions, rather like crystals in the metal, where the atoms all point the same way. In materials like iron these atoms have a tiny magnetic field caused by the moving electrons forming little electromagnets.

A key aspect of physics is that processes tend to be reversible. In this case we should be able to make electricity using a magnet. Well we can. In fact it is how pretty much all electricity we use is made (the only bit that isn't comes from solar panels).

Take a wire and move a magnet near it. You will generate a small electrical current. How do you make a larger current? You do the same trick as before and coil the wire up!

In 1862 Maxwell was able to rewrite all the existing equations for electricity and magnetism into a simple set of electromagnetic equations.

Rise and Fall of the Plum Pudding

One of the great all-time questions is 'what are things made of?'. Most answers have come down on the side of having a set of basic building blocks to make everything. Initially this was investigated by philosophers, then alchemists and then chemists. It took physicists to sort it out!

I'm sure you've heard of one of the earliest models we know about. Everything was made of some combination of four elements – earth, air, fire and water. This had the benefit of only having a few building blocks and had connections between properties and contents.

The next major model was the periodic table of the chemist Mendeleev. All matter was made up of atoms, with one type of atom per element (eg. iron, oxygen). Over 100 elements have been identified – a huge number for them to be basic building blocks!

Take a piece of gold. Keep cutting it into smaller pieces. It each lump will still be gold. Eventually, it was thought, you would get to a point where you couldn't cut it up any more. The Ancient Greek word 'atomos' (meaning 'uncut') was used to refer to such small lumps of an element. The word means 'indivisible' - these small lumps were thought to be fundamental.

However it was discovered by physicists that electrons came from inside atoms. That meant that atoms must be made of something else. The hunt was on.

In the early 20th century Rutherford studied the Plum Pudding Model of JJ Thompson. This held that the negative electrons were held in a positive 'batter'. He set some of his students, Geiger and Marsden, to test this.

In 1909 they fired alpha particles at gold foil. They recorded where they came out. The results surprised everyone. Most of the alpha particles went straight through. This showed that the gold was mostly made of empty space! Some were reflected through small angles. A few bounced almost straight back. This showed that there was something small and dense inside the gold.

From these observations the nuclear model of the atom was invented. It had the electrons orbiting a central nucleus, a bit like planets round the Sun.

Other scientists started to ask what the nucleus was made of. They studied it and discovered protons and neutrons. It was found that very atom of a particular element had the same number of protons. This meant that everything physical was made up of just three types of particle - protons, neutrons and electrons. That was very neat and beautiful. The physicists handed back to the chemists and let them get on with bothering moles.

Radioactivity

Radioactivity – Spiderman, the Hulk, Chernobyl, cancer. Scary subject? It can be dangerous but it can also be life-saving.

We are going to have to look at really small things now. In fact, for radioactivity we are interested only in the nucleus of an atom, that tiny bit in the middle.

Remember that the number of protons in the nucleus tells you what element it is – gold, iron, oxygen etc. For each element there can be a range of different numbers of neutrons; this makes different isotopes of the element. Normally one of these isotopes is stable, in other words it stays the same. We think of this as the normal isotope of that element. Other isotopes are unstable, in other words they can change.

When one of these unstable nuclei (the plural of nucleus) changes we say it has decayed. It decays by releasing another particle. This process is what we call radioactivity. The really cool bit is that it is random. We can give a very good prediction of how many nuclei will decay in a given time but we cannot predict when a particular one will decay. This isn't because we don't know enough. It is actually fundamental to the theory itself that the very nature of decay is unpredictable!

So, what are these particles that are emitted? There are three kinds. Alpha is the heaviest and is made of two protons and two neutrons. Beta is the next heaviest and is an electron. Gamma is the lightest and is a photon (a particle of light – think photograph).

The three types have different properties. Alpha, for instance, is stopped by as little as a sheet of paper whereas beta needs a thin sheet of metal to stop it and gamma will go through even a thick sheet of metal.

Radiation is all around us. Put a Geiger (yes, the same guy who proved the existence of the nucleus) counter in a room and it will be clicking away merrily as it detects the radiation. This is called the background radiation and is always there. It comes from space, from rocks, from the air, from nuclear bomb tests, even from us. It is higher in some places than in others, depending on what the local rocks are.

Exposure to radiation can cause damage to your body. When one of these particles hits atoms inside you it knocks electrons off. This forms things called ions and they can kill cells, mutate them or cause them to multiply out of control. There are many possible results of this but the one everyone will have heard of is cancer.

However, you will probably have also heard of radiotherapy. This is one of the tools used to treat cancer. It involves using radiation to kill cancer cells.

The Mass Defect

As studies into isotopes continued another decay was discovered. In this one the nucleus of a heavy atom splits into two lighter ones. This is called fission.

One of the things that made fission so appealing to scientists (and governments) was that whenever a nucleus split it created energy. Yes, you read that correctly. But surely energy cannot be created or destroyed? True, so it must have been being transformed from a type they didn't recognise into a type they did.

Well, it gets worse. When they were able to study the nuclei with really sensitive equipment they found out that the total mass at the end was less than the total mass at the start. The conservation of mass was as fundamental a principle as the conservation of energy. This difference in mass was so against the understanding of physics that it was called the 'mass defect'.

Eventually the problem was resolved. If you allow mass to be transformed into energy it makes sense. In a way mass turns out to be a form of energy.

Einstein came up with a very famous equation that links mass and energy – $E=mc^2$. This tells us how much energy is released when a certain mass is transformed.

Physicists don't like having exceptions in their laws. We weren't happy with the Principle of Conservation of Energy and the Principle of Conservation of Mass not being always true. So a new law came in to replace them – the Conservation of Mass-Energy.

It is as if mass and energy are two sides of a coin. There are always the same number of coins but they can be turned over so the numbers of heads and tails can change.

The Speed of Light

I am sure you'll have heard that nothing can go faster than the speed of light. But why?

In order to speed up an object needs to have a force applied (yep, that's Newton's 1st Law of Motion). The faster something moves the more kinetic energy it gains. As you'll remember from the last chapter this means it will gain mass. So, according to Newton's 2nd Law of Motion, it will take a larger force to accelerate it than before.

This leads to the idea that the faster something moves the harder it is to move it faster. It opens up the possibility that there is a speed above which it won't be able to accelerate.

A physicist called Lorentz had been playing around with some theoretical situations to see how the universe could behave. One of them was 'what if the speed of light in a vacuum was the same whoever was measuring it'. This lead to some cool equations, called the Lorentz Transforms. These say that as something gets faster it gains mass. Well, we already guessed that; but there's more. As it gets faster it gets shorter and time slows down for it!

If the object reaches the speed of light it will have infinite mass, no length and time will stop. If it were to somehow go faster than light it would have less than no mass, less than no length and time would go slower than not flowing.

In 1905 Einstein took the extra step of saying that the world actually did behave like this. He started with the Transforms and made predictions about the world. Time and again experiments have confirmed these predictions. For instance two atomic clocks were synchronised at an airport. One was flown around the world while the other remained at the airport. When they were reunited the one that had been travelling at speed was slow by exactly the amount predicted.

Warping Space-Time

Warp factor 5 Mr. Sulu. It's not actually that far-fetched. There are things that warp the surrounding universe and cause some very odd behavior.

In most situations Newton's Laws of Motion and his Law of Universal Gravitation work perfectly well. They allowed NASA to put men on the Moon and they accurately predict where all the planets will be any time in the future. All but one planet anyway. Mercury doesn't behave itself. Its orbit moves twice as fast as it should, according to Newton.

In 1916 Einstein published the General Theory of Relativity. It had occurred to him that gravity and acceleration were equivalent, in fact he called it the 'equivalence principle'. I am sure you have noticed that when a lift accelerates upwards you feel heavier, well that is what got him started. Imagine you are in a rocket with no windows. You feel heavy; is that because you are on a planet with a lot of gravity or is the rocket accelerating? There is no experiment that would be able to tell – all the laws of physics are the same regardless of whether you are accelerating or in a gravitational field.

His next step was to realise that space and time are not as different as they may first appear. An event is identified by a set of four coordinates. One of these must be time – if you want to meet someone it is no good just telling them where, you also have to tell them when. Einstein called this four-coordinate system 'space-time'.

Finally he explained that things with mass bent (warped) space-time. Gravity was just the slope of space-time – it was as if things were rolling down it. It is often describe as being like a rubber sheet with balls on it.

Now the equations that came out of this model didn't differ from Newton's by a measurable amount in most cases. This was good, as Newton's worked well. However Mercury is very close to the Sun and space-time is significantly warped there. In that situation Einstein differed from Newton; indeed exactly matching the observations.

The warping of space-time also predicted that a beam of light would be bent when it passed close to a large mass. In 1919 a British expedition was able to observe stars during a solar eclipse. Those whose light had to pass close to the Sun appeared out of position, confirming that the light from them had been bent.

As people started to play around with the theory it was soon realised that if enough mass were concentrated into a small enough space it would warp space-time so much that light shone upwards from the object would be bent back on itself; it would behave just like a ball thrown in the air. These objects were named 'black holes'.

As it always seems to when Einstein is involved the physics gets really weird. As we have said, mass warps time as well as space. That means that as the slope of space-time gets steeper time runs slower. A clock at sea level runs slower than a clock on top of a mountain.

If you were to head off in a spaceship and orbit close to a black hole time would slow down for you by a significant amount. When you came back you would have experienced a lot less time than those you left behind. You could have spent five years orbiting the black hole but the people at home would have spent ten years. You would have traveled five years forward in time!

Wave-Particle Duality

Some things we are happy to call matter – tables, chairs, squirrels. Some things we are happy to call waves – sound, water ripples, Mexicans. However some things have lead to long arguments about what they are. Light is a significant example.

Back at the end of the 17th century Newton believed that light was made up of little particles (he called them corpuscles) and so was matter. Huygens believed light was waves. At the time Huygens won the debate with a series of experiments demonstrating how light did indeed behave as a wave – it could spread out through a gap (diffract).

The matter (ha ha) seemed settled until we started being able to study the sub-atomic realm.

In 1905 Einstein published a paper on an obscure phenomenon called the 'photoelectric effect'. It had been observed that when light was shone on some charged metals they lost their charge. However it wasn't all metals or all types of light. Zinc, for instance, retains its charge when white light is shone on it but loses it when ultraviolet light (the type used in tanning booths) is shone on it. This couldn't be explained if light were a wave. Einstein realised that this was proof that Newton had been right and light was indeed made of particles. He called them 'quanta of light' or photons. The branch of physics called quantum mechanics was born.

But hang on, Huygens' experiments proved that light was a wave. Those experiments still work today. So what is going on; surely they can't both be right?

Well actually they are. It seems that light is both a wave and a particle. It just behaves differently depending on the circumstances. Sound familiar? Yes, it is just like mass-energy and space-time. In this case the principle is called 'wave-particle duality' and we say that light is made of wavicles (from WAVe-partICLES).

So, if we strongly believed that light was a wave but it turned out to be made of wavicles, what about things we strongly believed were particles?

In 1906 JJ Thompson received a Nobel Prize for proving that electrons were particles. He had done this by showing they had quantised mass and charge – they came in fixed lumps rather than being able to have any amount. In 1937 his son, George Thompson, received a Nobel Prize for proving that electrons were waves. Nowadays we recognise that they too are wavicles and both Thompsons were correct.

Big Bang Theory

There are two main classes of theory about how the universe began. One says it started, the other says it has existed forever. The debate between these has existed for a very long time indeed!

When Einstein finished his first attempt at the General Theory of Relativity he noticed that his equations meant the Universe was expanding. He thought this was ridiculous and decided his equations must be wrong. He added something called the Cosmological Constant which removed this problem.

In 1927 Lemaître proposed the Big Bang Theory. This postulated that the Universe began with a large explosion. This would mean that it would probably still be expanding. Various physicists noticed that this was in agreement with General Relativity's initial formulation, but there was still no proof.

However, in America Hubble was working on the problem. He was cataloguing the speed of galaxies and their distance from Earth. When his results were published they showed that nearly all galaxies were moving away from us and that the further away a galaxy was from us the faster it was moving away. This was taken as evidence of the Big Bang Theory. Einstein removed the Cosmological Constant from his equations and described it as "the biggest mistake of my life".

Hang on, all the galaxies are moving away from us? Doesn't that put us at the centre of the Universe? Well yes and no. We would see the same effect whichever galaxy we were in. All the galaxies are moving away from each other. We could equally well call any point the centre of the Universe.

This is all very well, but is there any direct proof of a Big Bang? Now that had to wait a few years for one of the greatest bits of experimental physics ever.

In the 1960s Penzias and Wilson were working on microwave reception. They had a large horn antenna and decided that they really should check to see whether their device correctly read zero. This is an important check for any piece of scientific equipment. They pointed the horn up into space away from all known sources of microwaves and took a reading. They found a systematic error. In other words they didn't get zero when they thought they should.

Over the years they tried everything they could to remove this error. They even employed hawks to prevent pigeons leaving their droppings in the horn. Eventually they concluded that it wasn't a systematic error after all, but a real signal from space that existed wherever they pointed the receiver. They published their findings.

Independently various people had been working on the idea that the Big Bang would have left behind a microwave signal. This was because the radiation from the explosion would have been stretched as the Universe expanded and ended up as microwaves. This is referred to as cosmic microwave background (CMB). One of these physicists was Dicke, also working in America.

A friend showed an unpublished papers based on Dicke's work to Penzias. He then realised the significance of what they had found. The three men finally published their two paper in the same edition of Astrophysical Journal Letters. Penzias and Wilson received a Nobel Prize for their work in 1978.

The Big Bang Theory now had much stronger evidence. Whilst it has undergone various modifications it is still at the heart of our understanding of the Universe.

The Standard Model

Scientists often stick with a good idea when they have found it. After all we have found time and again that one aspect of the world is often mirrored elsewhere.

The same technique as had been used by Rutherford's team was eventually turned on the protons and neutrons themselves. It turned out that protons and neutrons are not fundamental particles after all.

There are three particles inside each of them, called quarks. Protons have 2 up quarks and 1 down quark; neutrons have one up quark and two down quarks. Don't worry about the names – they are just names that are easy to remember.

OK, so there are still only three particles making things up – electrons, up quarks and down quarks. Still simple. Sadly it didn't remain that way for long. Other particles keep turning up!

Firstly there aren't just two types of quark, there are six. We call them flavours. Secondly there are six things called leptons (electrons and 5 others).

quarks		leptons	
up	down	electron	electron-neutrino
charm	strange	muon	mu-neutrino
top	bottom	taon	tao-neutrino

But that is not all. Every single one of these 12 particles also has an anti-particle!

These are used to build other particles. Ready for some more new names? Here we go…

Particles built of quarks are called hadrons. There are two kinds of hadrons. Baryons are made of three quarks (like protons and neutrons). Mesons are made of quark-antiquark pairs.

But that is not all. We can build all particles from these. What about light? Well that is made of photons, not too bad. What about forces? For these we need another class of particle, called bosons.

There are only four forces. Any force you can think of will be one of the four, in fact it is likely to be one of the first two of them. Electromagnetic forces are carried by photons; that covers pretty much everything as all contact forces are actually electromagnetic interactions between electrons! Gravitational forces are carried by gravitons (well possibly).

What about the other two forces then? One is the strong nuclear force that holds nuclei together. That is carried by gluons. These cause interactions between the different colour quarks (yes, quarks have a colour as well as a flavour). The other is the weak nuclear force which is involved in radioactivity. It is carried by W^+, W^- and Z^0 bosons. These change the flavour of quarks, in some cases leading to one element changing into another.

All together these particles form the Standard Model. On the face of it this is the physics version of stamp collecting.

Entanglement

This is where quantum mechanics starts getting really weird. In fact it is so weird that Einstein refused to accept it. He dedicated a sizeable chunk of his career to disproving it. It must have been a sore point for him that his Nobel Prize was awarded for the Photoelectric Effect which started the whole thing off!

The first step to seeing the problem is Schrödinger's Cat. This was a thought experiment that came out of an exchange of letters between Einstein and Schrödinger debating quantum mechanics.

In essence a cat is put in a box with a contraption designed with a 50:50 chance of killing the cat. Classically we would say that before we open the box the cat is either dead or alive and there is a 50:50 chance for each possibility. We just find out which when we look inside. Quantum mechanics, however, says that before we open the box the cat is both dead and alive in equal amounts. When we look inside we cause one outcome to happen (show off by saying the observer collapsed the wavefunction).

This is a significant change to how people doing experiments are viewed in science. There had always been the need to avoid the scientist affecting the results, by breathing on the apparatus for example. However now quantum mechanics was saying that however careful you are you actually trigger changes in the experiment simply by looking at it. But hang on, couldn't the cat look to check it was alive? What about an omniscient god who observes everything? What actually makes things collapse into a particular outcome?

The next step is to set things up so two things are tied up in the experiment, or entangled. Until one is observed the outcomes are mixed up. When one is observed the outcome of both is fixed. This would happen instantaneously however far apart they are. That required information to travel infinitely fast, significantly breaking the speed of light. Einstein called this 'spooky action at a distance'.

Experiments, on particles not cats, have been carried out to test this. Most significantly the Aspect experiment of 1981-2 proved that entanglement allowed information transfer faster than the speed of light. This opened the way to improved communications and even teleportation. In fact at the time of writing the record for teleporting photons stood at 89 miles and for teleporting matter at 21 metres.

There are times in science when a new theory comes along that makes a radical difference to how we see the world. When this happens, there is often an argument about whether it is describing how the world actually works or is a model that makes good predictions it. For instance the Catholic Church really liked Galileo's ideas as a model, because they helped predict Easter, but they objected to him saying it was how the world worked. For years the particle model of thermodynamics was not believed to be an accurate description of the world. Now quantum mechanics had brought the ultimate 'impossible' theory - it produced the best predictions of all theories but it seemed that the world couldn't work that way.

The greatest meeting of scientists the world has ever seen was convened at the Solvay Institute in Brussels to try to resolve this issue. It was attended by Einstein, Curie, Schrödinger, Planck, Pauli, Heisenberg, Bragg, Dirac, de Broglie, Bohr and Lorentz to name but a few. The majority came down on the side of those saying that describing how the world worked was less important than making predictions.

Strings and Branes

Einstein's equations of general relativity were expressed in four dimensions. They were able to account for all gravitational effects as being ripples (fluctuations) in space-time.

But there are four forces. I am sure by now you have noticed that the biggest sign of progress in physics is spotting connections and combining things. So, where do we go now?

In order to make general relativity work Einstein had had to use four dimensions. People quickly wondered what would happen if a fifth dimension was used. Well the answer was better than they had imagined. Using the equations for general relativity but expressing them in five dimensions they got the normal gravitational properties and also Maxwell's equations for electromagnetism. In other words using five dimensions combined gravity and electromagnetism!

Flushed with this success physicists started trying larger numbers of dimensions. They wanted to find a single system that would include all four forces. The hunt for a grand unified theory was well and truly on.

Sadly adding more dimensions didn't produce anything helpful at first. The base equations needed tweaking.

The first strong contender was superstring theory. This has either 10 or 26 dimensions all coiled up into a string. The second was brane theory where the dimensions are stretched into sheets (membranes). In both cases the hope is that all mass, energy and forces can be explained as fluctuations of the string or brane.

Home Experiments

Some experiments you can do at home.

Phonebook Friction

This simple experiment demonstrates how strong friction can be.

<u>What You Need</u>
phonebook x 2

<u>Instructions</u>
Take your books and open the first page. Put one on top of the other so the spines are facing away. Alternately open a page from each book so they build up the pile. Eventually all the pages will be interleaved (if you use smaller books this will take less time but don't do it with anything you don't want to damage).

Take a firm hold of the spines and try to pull the book apart. You may want to get someone else to try to pull too.

<u>Comment</u>
Friction is the force that tries to stop surfaces rubbing past each other. The larger the contact area the larger the friction. In this example the area of pages in contact is immense, over $15m^2$. This means that the friction is huge and it would take a very large force to pull the books apart.

There is a trick though. Whilst pulling gently, blow from the side to separate the pages. This could form the basis of a pub trick!

Liquid Fibre Optics

This experiment demonstrates light being bent round a stream of water.

What You Need
see-through bottle

dark paper or card

torch

hammer

nail

Instructions
Make two holes in the cap of the bottle using a hammer and nail. Fill the bottle with water. Wrap the paper round the bottle leaving the same length again beyond the bottom.

Switch the torch on and put it in the paper cylinder below the bottle. Switch the lights off and slowly pour the water into the sink with the holes one above the other. Try running your fingers in the stream of water.

Comment
The light from the torch enters the water in the bottle. Some of it goes out of the hole and along the stream of water. When it hits the edge of the water it does so at a large angle. This means it reflects (totally internally reflects to be precise). It does this all the way along the stream even though it is bent.

This is how optical fibres work for broadband and Christmas decorations.

Air Cannon

This simple experiment demonstrates how to make vortices that travel a long way.

What You Need
pringles tube

balloon

elastic band

nail

hammer

scissors

candle

Instructions
Remove the lid from the tube. Cut along the length of a balloon and fasten it over the open end of the tube using the elastic band.

Make a hole in the other end using the nail and then cut a circle out using the scissors.

Light the candle. From a metre away point the tube at the flame and twang the rubber on the end. Once you have got the aim right try from further away.

Comment
The twanged balloon causes a pressure build up in the tube. Air is forced out of the end. This curls around the edges of the hole and becomes a circular vortex travelling forwards. Like a tornado the air is moving very fast around but keeps together as is moves along.

Freefall Trajectories

This neat experiment demonstrates that horizontal speed has no affect on how fast an object falls.

What You Need
ruler

marble or coin x 2

Instructions
Place the ruler diagonally over the edge of a table. Place one coin on the overhanging end of the ruler. Place the other on the table in front of the ruler. This works best on a hard floor and you will need a decent bit of space in front of the table.

Swiftly hit the end of the ruler that is on the table outwards. One coin will drop and the other will be flung across the room.

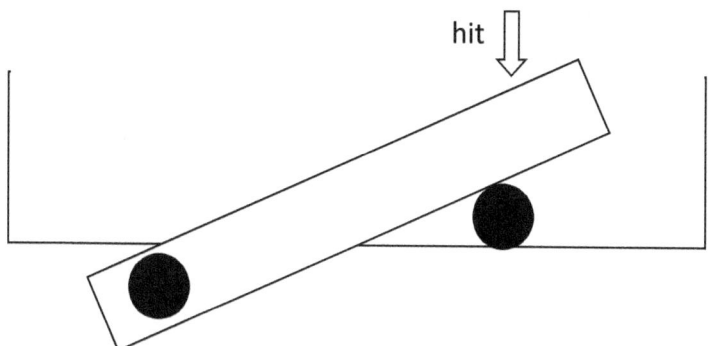

Comment
Horizontal and vertical components of motion are independent. Both coins are dropped from pretty much the same height so should hit the ground pretty much together even though one has a large horizontal speed (velocity would be an even better word).

Electrostatics

This simple experiment demonstrates how like charges behave.

What You Need
balloon x 2

thread

wooly jumper

Instructions
Blow up two balloons. Tie a thread onto each one. Rub them both on your jumper for a minute or so.

Hold them up by the thread and try to bring them together.

Comment
Rubbing the balloons on your jumper strips electrons off the balloon and onto the jumper. This means that both balloons are positively charged.

Like charges repel and opposites attract. As both balloons have the same charge they repel and try to stay apart.

Bending Water

This simple experiment demonstrates how a neutral thing can be attracted to a charged object.

What You Need

plastic comb or ruler

woolen or fur cloth

Instructions

Turn on a tap to a gentle flow which doesn't quite break into droplets.

Rub a plastic rod, like a comb or ruler, with a natural piece of fabric - wool, fur or cotton.

Bring the rod near the water.

Comment

Rubbing a plastic object with a cloth strips electrons off one onto the other. This leaves them charged - the one that lost electrons is positive and the one that gained them is negative.

When it is brought near the neutral water it causes the charges inside to split and the water is attracted to the rod.

Electromagnet

This simple experiment demonstrates how electricity can create a magnetic field.

<u>What You Need</u>

battery

wire

iron nail

paperclips

<u>Instructions</u>

Wind the insulated wire round the iron nail as many times as you can. Connect the bare ends of the wire to a battery. Be careful as the wires could get hot.

See if you can pick up any paper clips with the end of the nail.

<u>Comment</u>

A moving current generates a magnetic field. This is fundamental to making motors and many other electrical devices.

Drinks Can Rocket Motor

This simple experiment demonstrates how a rocket motor works.

What You Need

drinks can

nail

thread

Instructions

Punch a set of holes round the base of the drinks can, using the nail. When the nail is in each hole, push it to the left to make the hole point that way.

Tie the thread onto the top of the can to make a long handle. Fill the can with water while holding the thread.

Comment

The water is pushed out of the holes by the water above it. In turn, by Newton's 3rd Law, it pushes back on the water still in the can, causing it to spin.

Rockets work in the same way as one of the hole, only they push gas out the back by heating it. The stuff thrown out of a rocket is called reaction mass.

Inertia

This simple experiment demonstrates the inertia in objects.

<u>What You Need</u>
cup
playing card
coin

<u>Instructions</u>
Place a playing card on top of a cup and a coin on top of the card. Flick the card hard horizontally.

<u>Comment</u>
All objects with mass have inertia. This means that it requires a force applied over time to move it.

There is friction between the card and the coin so if the card were pulled slowly the coin would come with it. However, flicking the card quickly means the coin is left behind.

If you are feeling brave you could try the tablecloth trick - yanking a tablecloth out from a table so fast that the plates, glasses etc. are left standing!

Bernoulli Paper

This simple experiment demonstrates that fast moving air has less pressure than slow moving air.

<u>What You Need</u>
a sheet of paper

<u>Instructions</u>
Hold the paper by two corners of the short edge. Make sure it starts off horizontal before it droops down.

Place the edge below your bottom lip. Blow hard across the sheet.

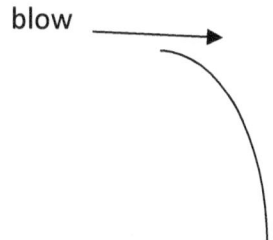

blow

<u>Comment</u>
Blowing makes the air above the sheet move faster than that below it. The faster moving air has a lower pressure than the slow moving air. This means that the air below pushes up on the sheet and it rises. This is called the Bernoulli Effect and is part of how an aircraft flies.

Superposition

This simple experiment demonstrates that waves can pass through each other.

<u>What You Need</u>
washing up bowl or bath

<u>Instructions</u>
Fill a bowl or bath with water and leave it to go still. Dip two fingers in and out some distance apart. Watch how the waves pass through each other.

<u>Comment</u>
Waves carry energy. The matter just vibrates up and down, it doesn't move along. This means two waves can pass through each other and carry on unaffected.

In the area where they do overlap they can interfere with each other, a process also called superposition. If two peaks meet you get a double height wave (constructive interference). If a peak meets a trough you get flat water (destructive interference).

Seeing Sound

This simple experiment demonstrates that sound is a series of vibrations.

What You Need

drinks can

scissors

balloon

string

glue

small mirror piece

torch

Instructions

Cut the top off the can. Be very careful as this will be sharp. Cut a balloon down the middle and stretch it over the cut end. Tie it on.

Glue the piece of mirror to the middle of the balloon. You can get little mirror tiles from craft shops.

Shine a bright torch on the mirror or sit in bright sunlight. Find the reflection on the wall or ceiling. Place the other end of the can in front of your lips and talk. Watch the reflection.

Comment

Sound is a series of vibrations. Talking at the can causes it, and thus the mirror, to vibrate. The moving mirror reflects light into a moving pattern.

Spoon Reflections

This simple experiment demonstrates how curved mirrors affect light.

<u>What You Need</u>
shiny spoon

card

scissors

<u>Instructions</u>
Cut a set of small slits in the end of the card. Place it vertically on a table and place a torch behind it.

On the other side put a shiny spoon. Move the spoon around until the reflected light hits the paper. Try the bottom and the top pointing towards the light.

<u>Comment</u>
Light always reflects off a mirror at the same angle it hits it. If that mirror is curved the rays will be reflected in different directions.

The bottom of a spoon is a convex mirror. This makes light rays spread out (diverge). The top is a concave mirror. This makes light rays focus (converge).

Appearing Coin

This simple experiment demonstrates refraction.

What You Need
cup x 2

coin

Instructions
Place a coin in the bottom of a cup. Fill the other cup (or a jug) with water.

Position yourself so the coin has just disappeared behind the rim of the cup. Slowly pour the water in.

Comment
Light travels in straight lines. When the coin is behind the rim of the cup there is no straight line path to your eyes.

However, light changes direction (refracts) when passing from water to air. This can cause it to reach your eyes and you see the coin.

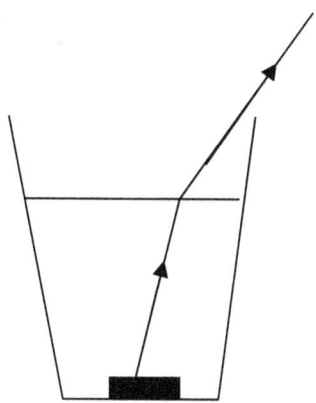

Pinhole Camera

This simple experiment demonstrates how to make a simple camera.

What You Need
shoebox or pringles tube

greaseproof paper

pencil or nail

glue

Instructions
Take the box and cut one of the ends off (if using the tube, simply remove the lid). Glue greaseproof paper over this end.

Using the pencil or nail make a small hole in the other end.

Point the end with the hole towards something and look at the greaseproof paper. Try looking at a light bulb or something else bright to start off with. If that is still too dim make the hole a little bit bigger.

Move backwards and forwards until you get a clear image.

Comment
Light rays head out from the object in all directions. By only using the ones that can pass through a small hole we are able to form an image on the screen.

It is upside down and back to front because the rays crossed over in the hole.

Spectrum

This simple experiment demonstrates that white light is made up of 6 different colours.

<u>What You Need</u>
bowl

mirror

<u>Instructions</u>
Fill the bowl with water and prop the mirror inside at an angle. Place in bright sunlight. Where the light is reflected will be a rainbow.

<u>Comment</u>
The wedge of water above the mirror acts as a prism. Light is refracted (changes direction) on entering and leaving the water. The red is refracted the least and violet the most (red has the longest wavelength and violet the shortest). This splits the light up into its colours, called the spectrum.

Contrary to popular belief there are only six colours in a rainbow. Newton wanted 7 colours so he inserted indigo. It is possible he did this to match it up to the 7 gaps in a musical octave. It is also possible it was because of his stout Christian faith and 7 being a good Biblical number for what he thought of as God's light.

Blue Sky

This simple experiment demonstrates why the sky is blue and sunsets are red.

What You Need
glass

milk powder

torch

Instructions
Fill a glass with water. It is even better if you can use a jug or even a tank. Shine a torch through. Nothing much will happen to the colour of the light.

Next add milk powder to the water and stir. Look at the torch through the water and into the water from above.

Keep adding more milk powder and repeating the observations.

Comment
The milk particles scatter some of the light. They scatter blue light more than red light as it has a shorter wavelength. This means that viewing from above you see blue light but from straight on you are left with red light.

Particles in the air do the same. The more are the light has to pass through the more gets scattered; sunsets are redder than midday sun. The more particles in the air the more light is scattered; sunsets are more spectacular in polluted air!

Compass

This simple experiment demonstrates a few bits of physics. Firstly it shows how to make a magnet, secondly the power of surface tension and finally the effect of the Earth's magnetic field.

What You Need

steel paperclip

magnet

cup

Instructions

Fill the cup with water.

Stroke the paperclip with the magnet a hundred times. Always use the same end and stroke in the same direction.

When you have finished that very gently lower the paperclip flat onto the water. It may take a few tries to get to right but you will be able to get it to sit on top of the water.

Notice which way the paper clip points. Try rotating the cup. If you are gentle enough not to sink it notice which way it ends up pointing.

Comment

Iron and steel are made up of tiny little magnetic bits called domains. Normally these are all jumbled up and cancel each other out. Stroking it with a magnet lines them up - a bit like combing unruly hair. This makes a magnet.

The surface of water is like a sheet of rubber due to the surface tension. It takes a reasonable amount of force to break through it.

The Earth's core generates a magnetic field. If a magnet is free to move it will align itself with this field. This is how a compass works.

Hovering Ball

This simple experiment demonstrates how a wing works.

<u>What You Need</u>
bendy straw

ping pong ball

scissors

<u>Instructions</u>
Take the straw and cut four slits into the short end, each about a centimetre long. Fold them out to form a stand.

Bend the straw into a right angle. Hold the straw with the long end horizontal and the stand upwards. Place the ping pong ball into the stand and blow into the other end.

Once you have got the ball stable in the air try rotating the straw so the stand is no longer underneath the ball. You will need to keep blowing!

<u>Comment</u>
Blowing on a ball to lift it should be no surprise. The air slows down on hitting the ball and applies a force on it - Newton's 2nd and 3rd Laws.

However the second part may be more of a surprise. As the ball starts to drop the air moves faster over the top than the bottom. The faster moving air has a lower pressure. This effectively sucks the ball back up.

Egg in a Bottle

This is great fun but often mis-explained.

<u>What You Need</u>
hard boiled egg

milk bottle

candle

matches

<u>Instructions</u>
Peel the shell off the egg.

Light the candle and drop it into the empty milk bottle. Check it is still alight.

Place the egg on the open top of the bottle. Wait and watch the egg get 'sucked' into the bottle.

<u>Comment</u>
The egg isn't sucked in. It is pushed in by the air outside because the air pressure inside has dropped. The reason for this drop is often incorrectly given.

It is not because the candle uses up the oxygen. It is true that the candle use some of the oxygen in the air to burn. It does go out because there is no longer enough oxygen compared to carbon dioxide in the bottle. However if this were the only thing going on the difference in pressure would be incredibly slight.

Actually, the candle is also heating the air. This causes it to expand. As the egg is just sitting on the top air can flow out past it and the air inside stays at room pressure. However once the candle is out the air cools. The egg forms a seal so the pressure inside drops.

Notice that the egg only goes in after the candle goes out!

Topping up a Bath

This simple experiment demonstrates that hot water floats on top of cold water.

<u>What You Need</u>
a bath

<u>Instructions</u>
Get in a cool bath (perhaps you could get in a hot bath then wait until it cools). Carefully turn on the hot tap and lie back disturbing the water as little as possible. Notice where the hot water goes.

<u>Comment</u>
When water warms up is expands. This means it is less dense; the same mass of water takes up more space. That means warm water floats on top of cold water.

Speed of Light

This great experiment actually lets you measure the speed of light!

What You Need

a microwave oven

butter

ruler

Instructions

Find the panel on the microwave. It should give the frequency of the waves. This will be followed by Hz. If it is followed by kHz you have to multiply by 1000. If it is followed by MHz you have to multiply by 1000000. Write this down.

Next take a pat of butter and chop it up into 1cm thick slices. Use these to cover a microwaveable plate.

Put the butter into the microwave and turn on for a couple of seconds. Look for melted bits. If these aren't clear put it in for another couple of seconds.

Measure between the melted bits. Write this down in metres.

Wavespeed is wavelength x frequency. If you multiply your two numbers together you should get something like the speed of light.

Comment

Microwaves are a form of light. They travel at close to 300000000m/s in air.

Mixing Light

The primary colours of light are not red, yellow and blue.

What You Need

torch x 3

white paper

variety of colours of see-through sweet/chocolate wrappers

elastic band x 3

Instructions

Fasten a different coloured wrapper to the front of each torch using the elastic bands. Turn off the lights and shine the lights onto the paper. See what happens when they overlap.

Try different colours of wrapper.

Comment

You should find that to make white light you needed red, green and blue.

Which two of these do you need to mix to make yellow?

Convection

This simple experiment demonstrates convection currents in water.

<u>What You Need</u>
drinks bottle x 2

kettle

food colouring

paper

<u>Instructions</u>
Fill one bottle with cold water right to the very top. Place a piece of paper on it so the paper gets wet. Put some food colouring in the other bottle and fill it with warm water from a kettle.

Carefully turn the cold bottle upside down (the paper should hold the water in) and place it on top of the hot one. Gently slide the paper out and watch the coloured and colourless waters mix.

<u>Comment</u>
When water warms up is expands. This means it is less dense; the same mass of water takes up more space. That means warm water floats on top of cold water.

Carrying On

If you want to do more reading or get some kit to try more elaborate experiments you ought to check out my online science store at http://astore.amazon.co.uk/sciencestuff-21...

About the Author

I studied at the University of Cambridge, leaving with a BA in Natural Sciences and an MSci in Experimental and Theoretical Physics. I went on to earn a PGCE specialising in Science and Physics from the University of Bangor. A secondary teacher for over ten years I have plenty of experience communicating scientific ideas.

I grew up in Lancashire, within easy reach of the Yorkshire Dales, Pennines, Lake District and Snowdonia. After stints living in Cambridge, North Wales and the Cotswolds I have lived in Somerset since 2002. I have been climbing, mountaineering, caving, kayaking and skiing as long as I can remember. Growing up I spent most of my spare time in the hills.

Landscape archaeology has always been one of my interests; when you spend a long time in the outdoors you start noticing things and wondering how they came to be there. At university I chose geophysics as one of my options.

I am an experienced mountain and cave leader, holding a range of qualifications including ML, SPA and LCL. I am also a course director for climbing and navigation award schemes.

My personal website, listing my work and interests, can be found at http://www.alasdairshaw.co.uk.

To receive email notifications of new physics publications please sign up to my mailing list at http://eepurl.com/bu7lun.

Other Publications

Physics Problems for GCSE
http://www.archaeoroutes.co.uk/edphys/problems.php

The Best Bits of Physics: Teachers' Edition
http://www.archaeoroutes.co.uk/edphys/readers.php

AQA GCSE Physics Revision Guides
http://www.archaeoroutes.co.uk/edphys/revision.php

AQA A-Level Physics Practice Tests
http://www.archaeoroutes.co.uk/edphys/exams.php

Independence (science fiction)
http://lrd.to/xZJGd4yPXo
Walking Through the Past (walking guidebooks to archaeological sites in Britain's hills and moors)
http://www.archaeoroutes.co.uk

www.ingramcontent.com/pod-product-compliance
Lightning Source LLC
Chambersburg PA
CBHW051246170526
45165CB00004B/1599